ALL ABOUT US

D1319646

by Bettina Sulzer Milliken

Creative Designs, Inc.
Albuquerque, New Mexico

Design and layout by Jonathan J. Morningstar
Edited by Doug Lindsey

Published by
Creative Designs, Inc.
11024 Montgomery NE, Suite 311
Albuquerque, New Mexico 87111
(505) 275-3030

Library of Congress Catalog Card Number: 97-66340

ISBN 1-880047-46-2

"Animals have their own world of which we can only visit. If we wish to attain their friendship, we must come to know the essence of their nature."

(unknown)

Dedication

I would like to dedicate this book to my close friends, the Orangutans, and Regina Frey and Veronika von Stockar, who taught me my love for animals. I have spent the most memorable days of my life in the rain forest with them. My gratitude is permanent. My love and respect for animals is deep.

Why I wrote this book

In September of 1975, I had the unique opportunity to visit a childhood friend of mine from Switzerland, Regina Frey, who was a graduate of zoology of the University of Zurich. She and her partner, Monika Borner, were the project directors for the Bohorok Orangutan Rehabilitation Station in Northern Sumatra, Indonesia. They constructed and began the station in 1972.

The station was in the Gunung Leuser Reserve, a vast area of protected rain forest and surrounded by rivers. Orangutans cannot swim, so they were naturally contained. The project was funded by the World Wild Life Fund and the Frankfurter Zoological Society.

The purpose of the station was to protect and rehabilitate Orangutans, whose life in the wild was being endangered by mankind interfering in their habitat. They had been pronounced an endangered species.

The first step was to convince the Indonesian Government to declare the captivity of or any dealings with Orangutans as illegal.

Any ape found could be officially confiscated and brought to the station. There they were first put into a quarantine for the necessary time to ensure that the animals were rid of any illness or bacteria which could be spread to the wild population or the free living rehabilitants.

After their release from quarantine they lived with the other rehabilitants out in the forest under supervision of the zoologists. They were fed twice a day with milk containing vitamins and bananas. This nutrition would get them fit enough to provide for themselves. Some left the station on their own, some had to be transported further away.

I spent three months there and it was undoubtedly the most important experience in my life regarding nature at its best.

With the following text and with special photographs, I would like to introduce to you WHO lives in the Sumatran rain forest. This book is written from the Orangutans' point of view.

The way I got to know the Orangutan in his forest is a lasting inspiration and discovery to me as to how regal, peaceful, remarkably intelligent and charming such a primate really is. We humans know little of these creatures and the environment in which they live.

Facts About Us

Where do we live? In Southeast Asia on Indonesian and Malaysian Borneo and Northern Sumatra.

How do we grow? Pretty much like humans; we are infants till 4, children up to age 7 and teenagers from then to 12 years old. After that we are basically mature.

How long do we live? Oh... till about age 36, 40! No one really knows!

How big do we get? Male Orangutans usually grow bigger than females. We can measure up to 5 feet or more and weigh up to 230 pounds.

How are we different from humans? We predominantly live on trees and travel by swinging from branch to branch. Our hands and feet are built for that function with long toes and short thumbs which work like hooks. We do not walk upright but use our fists for support when on the ground.

What do we eat? We are mostly vegetarians and eat fruit, leaves and flowers. But we do like insects and ants and sometimes even venture out to try bird eggs.

How do we live? We are awake during the day and sleep at night. This makes us *diurnal*. We do not live in groups and are quite solitary. We are very intelligent. But we are an **endangered species**.

OUR HOME

Far away in the rain forests of Southeast Asia we, the Orangutans, live. There are only two islands which we inhabit, Sumatra in Indonesia, and Borneo, which is part Indonesian, part Malaysian. The human beings, who live nearby, call us "Person of the Forest", "Orang" meaning person and "Utan" meaning forest in their language.

You will notice a difference between the Bornean and Sumatran Orangutans. The Orangutan from Borneo has a much darker, almost black face, and the male has noticeably large cheekpads. We from Sumatra are lighter in color, have smaller cheekpads and have a ginger colored mustache.

We are apes, great apes, NOT monkeys. The distinction is that we are bigger in size as well as having a more complex brain and we do not really have a visible tail. We are very closely related to human beings.

We live very peacefully, way up in the trees, and normally bother nobody as long as we have our necessary space to live in. Generally we mean no harm and we hang out so far up in the trees that humans cannot easily see us. Our family and friends is what we mostly care about.

Unfortunately, more and more human beings have come our way into the forest. They seem to like the wood of our rain forest trees a lot and have begun to cut them down as they see fit. Apparently they can sell the wood for a lot of money to be made into fancy furniture, floors and whatever else they like to use it for.

Besides gathering the wood, they have also begun to clear large areas of forest because they need to build huge buildings for rubber plantations, oil and gas sites and roads on which they can travel with their motorcars. All of a sudden there were more and more people in our habitat and along with them came a lot of terrible noise. We and all other animals in the rain forest did not like this one bit.

The human beings did not like us very much either. We believe they were frightened of our appearance and size. We were a threat to them. So when they decided to construct their buildings right in the midst of the trees in which we lived, they took out their guns and shot some of us big ones and captured others. A horrible time began. Our big and beautiful mothers were killed, and the people took the babies and the little ones.   God only knows what they did with them! Some were kept as pets in their houses, some were sold to zoos and some were given to medical laboratories where scientists would practice medical experiments on us! Can you imagine? Plus there was a trade exporting us beautiful, exotic creatures of our rain forest to foreign lands.

A difficult time began. More of our forest was being cut down and our life was seriously jeopardized.

Luckily, some human beings became aware of our plight. Zoologists came to our rescue. They started to negotiate with the government and were then allowed to find some undisturbed land in the rain forest. This was protected by state law and no deforestation was being permitted. So that is where the zoologists began to build the Orangutan Rehabilitation Station.

Rehabilitation means to teach captured Orangutans to live free and independently in the wild again. The people built a cute house for themselves and a house for some local helpers as well as cages for us in which we would be prepared for life in the wild. The station was constructed along the shore of a clean and re-freshing river, the Bohorok, which provided their water needs. The entire chosen area of rain forest was surrounded by rivers and, since we cannot swim, we were totally safe and could not get into trouble on the other side across the water. Human beings were not allowed to build on or intrude into our land.

The zoologists had to help enforce the government's already existing law which declared the keeping of Orangutans as pets illegal. A group of officials was formed, which authorized to confiscate any Orangutans found in captivity. That is when we were brought to the station where our new life and the people's work began.

The zoologists' house

The rain forest exists in different areas of the world. It is called rain forest because the uniquely numerous huge trees and flowering plants absorb so much moisture that it is the spot where it rains the most in the world. Because of all the rain many wonderful luscious plants, creatures and majestic trees grow there. This deep green and humid part of the world creates humidity in the air and in turn makes rain. The jungle is the most fertile place on Earth, which is why the rain forest is vitally important for the Earth and life. It is a source of water that has been naturally built into the climate of the globe.

Whatever falls to the ground in the forest decomposes rapidly and provides fertilizer for the plants. It circulates within the plants and trees which, due to their large size and huge quantity, keep the soil fertile. Once that organic material is destroyed by deforestation, the soil will be washed away by the plentiful rainfall. When the roots are dead the earth turns into mud and nothing grows anymore. Without humidity in the air, there is no rain, no water, no growth. The sun is so hot without clouds and rain that none of us could survive. The Earth would all turn into desert.

The destruction of the rain forest does not only affect our lives, but human beings could not survive either.

Mother Nature has been carefully balanced since her origin. It is unwise for human beings to disrupt this delicate and magnificent creation.

To save our rain forest is one of the most pressing issues in our time.

A glimpse of the glorious Rain Forest

Remember:

1) *The Importance of the Rain Forest*
2) *Stop cutting down the Rain Forest*
3) *Our home, the Orangutan Rehab Station*

ME, MO

Hi! This is me, Mo. I am a female and a bit older than a year and came to the station as an orphan.

Humans shot my Mom so I did not get to be raised by her for my first four years as is normal for us Orangutan babies. My Papa, like all Orangutan Papas, do their own thing, cruising through the forest. He met my Mom, was with her for a bit, and then moved on. That is how they like it.

My Mom was so beautiful. She was the best climber you have ever seen and she would swing from tree to tree and I just held on to her long, strong red coat. Every so often she would stop for me to nurse a bit. When I got older she began to show me how to find and prepare all the delicious fruit growing in the rain forest. It is the sweetest and juiciest fruit you'll ever know.

Eventually I would let go of her belly and she would begin to teach me how to climb. It is hard to learn, but also fun. It is just wonderful to be an Orangutan because we have very long arms with hands that have long fingers and a strong thumb. So we can just swing from tree to tree with great speed. Then of course we have strong legs and what is really special about us, is that we have feet which work just like hands, with long toes and we can hold anything with our feet. My Mama told me that human beings only use their feet to stand on and walk with but do not really know how to hold anything with their toes. I loved my Mama, because she was such a skillful teacher.

Every evening, after the sun went down, we would climb up to the crown of the tree and prepare a nest for us to sleep in. Why we went way up is because we did not want the clouded leopard or the tiger, who also inhabits the forest, to hunt us during the night. They do not like to climb up that high. Mama would interlace twigs and leaves into suitable branches and in the end our sleeping nest looked just like a hammock that you humans use. They were so comfortable! Falling asleep, high up in the canopy of the trees, being well assured that no one would harm us, was quite heavenly.

I was happy to be alive in such a special place. One day, we were sitting in a tree eating the ripest and most delicately flavored durian fruit, our favorite, when we heard noises of men coming. It never really meant good things and we became alert. All of a sudden we heard their voices getting very excited and it occurred to us that they had detected us. Well, before my Ma or I had any time to flee, the worst imaginable thing happened. The men took out their guns and killed my Mama.

So here I sat, up on the tree, my mother dead, all alone and very sad. The people climbed up the tree, rather well, I may say, and launched a net over me and took me down. I fought a bit, but as you know, I was not an accomplished climber and therefore was easy prey.

I was miserable wondering where I would end. They took me into one of their motor cars and off we zoomed into town.

There I was put into a box in a noisy and hot house. Every day someone came and took care of me. They were nice, and gentle with me, but I missed my Mama. I wanted to go back to the forest.

Actually it was not before too long when a group of men took me in my box into a car. We drove for a while. Then two other men carrying a long pole lifted the box out of the car and hung me in the box on the pole! Where were we going? Swaying back and forth, I realized that the men were carrying me along a jungle path. A delightful feeling came over me! It felt so good to be back in the forest, even in a cage. We got to the river where they put my box and me into a tiny ferry boat and we crossed the water. After reaching the shore the men walked a little further to a pleasant house in the middle of the rain forest. The women at the house took me out of the box and gave me a warm welcome.

I had reached the Orangutan Rehabilitation Station. It was a blessing!

There were a whole group of other Orangutans around. I was kept inside the people's house in yet another box, but this one was blanketed. The elder group was outside in the forest and were fed twice a day. They were given bananas and fortified milk to make them resistant and healthy.

It was just fine for me to be with the women. They were human Mamas and treated me accordingly. Every morning when they got up I was taken out of my sleepbox and sat with them at the table for their so called breakfast. We Orangutans are vegetarians; we love all delectable fruit and also like to try new foods. The people ate soft boiled eggs and after my first try I wanted nothing else but soft boiled eggs. I used to throw a fit every morning until they realized that I was serious about my weakness for eggs. After that I got my soft boiled egg every morning.

I would sit in a chair and munch away, and for the first time since my Mama was shot, I felt loved. I guess I was pretty lucky to have found the zoologists.

They were busy people. They had to cook their food and continually made notes in books. With them also lived a little black and white dog named Wumi who scared me a bit. Wumi was curious and loved to hunt. She would sniff me up and down and I never had the experience of a black, cold, wet dog nose so close to me. But it all kept me intrigued about my new found life.

Most days one of the women would take me out and give me a lesson in tree climbing. When I watched them stumble up the tree, I remembered my Mama who was so much better equipped to climb than they were. However, it was good for me to be taught, because having seen the other apes, I knew that I wanted to become the best climber in the whole rain forest and hopefully as elegant as my mother. They had treats ready for me; so when I performed to their liking I was given beans which I really enjoyed.

I was happy at the station and grew bigger and strong. The zoologists began to think that perhaps one of the older females at the station would want to adopt me. They started to bring me up to the feedings in the afternoon. Everybody was curious about me and came to meet me. It was a little overwhelming and I did not care for all these introductions. Some of them wanted to pick me up and take me into the trees with them! I refused loudly. I made these really nagging, whiny noises, and clung onto my human Mom. She took me back to the house and I continued living with the zoologists until I became really independent.

And one day, I took off into the forest to live the free and peaceful life of a mature Orangutan lady. The rehabilitation station surely was a good place for me to grow up and I hope that these really caring people will teach countries, governments and human beings to let our rain forest and US be!

Remember:

- *Mo remembers her Mom*
- *Mo meets Man*
- *Mo moves to the Rehab Station*

HANGING AROUND

The zoologists built some cages up above their house, about a twenty minute walk on a steep jungle path up the mountain, where we would all "hang out." The cages were there for the bigger young ones, about five years of age, who still had to be sheltered during the night. On top of these cages was all the action of the rehabilitation station.

Twice a day, early in the morning and in the late afternoon, the people's helpers carried up lots of sweet, tender bananas and milk for us. The zoologists decided that we needed to get all the necessary vitamins for our health and survival in the wild. We loved that stuff; out there we do not come across delicious and creamy white milk very often! So we would all scurry onto the roofs of the cages immediately in order to get our fair share of it.

The people came every afternoon too. They wanted to observe and study us during the so called teaparty.

We could not wait for that teaparty. It was also fun for us to study and observe the human beings. We always had good fun together. Sometimes when the people were late, we would start down the path towards their house, but they did not like that at all. The zoologists were strictly not allowing us to come down toward their houses; remember? We were supposed to be undomesticated! They wanted us to get accustomed to the undisturbed rain forest.

We would always greet them with overjoyed squealing noises and big hugs that the people just loved. Sometimes we gave them loud, juicy kisses which they also enjoyed. Although they discouraged us from doing this, since we were to live independently, there was a mutual affection between us.

After the welcoming ceremony we rushed to grab for the milk cup and all the bananas we could possibly hold in our hands and feet. The problem was that there were about twenty-two of us trying to get that one milk cup and we often began to fight over it. Sometimes one of us had a real temper and began to throw the milk cup around and the whole feeding became a very messy affair. That angered the people and they would reprimand us. Many of us ended up having more milk on our bodies than in our mouths!

Feeling full and nourished, we then very happily relaxed a little, lying on our back looking up at the trees or daydreaming. We also loved playing with the people. We are so curious and wanted to experience everything hands on! Not all Orangutans were so outgoing; some just stayed to themselves, perhaps thinking about their Moms.

Occasionally, some of the bigger Orangutans, who had already left the station and lived on their own in the forest, would show up again for the feeding. It was irresistible knowing a place where there was a daily supply of bananas. Mehra, a beautiful, full grown female, checked in every so often. She sometimes brought along a wild friend whom she met in the forest. He would sit on a tree high up above the cages, watching, while Mehra came down onto the roofs and got bananas for the two of them. She then carried them up to him and the two thoroughly savored the feast. They would hang out a little and then continue on their path.

Mehra was always nice to us, and we were happy to know her. It was also encouraging for us to see how beautiful and strong we could get and were looking forward to the day when we would be cruising through the rain forest again.

Wumi waiting, shortly before the teaparty.

This is Mehra's friend, the big, wild, male Orangutan. He never came close to us, but was always very friendly from afar. A full size male Orangutan is easily up to five feet tall and 230 pounds in weight. Now that is big! Females are smaller.

We loved his power, grace and politeness and were seriously wondering what he thought about this discovery in the middle of the remote rain forest. Human beings, fresh bananas every day *and* milk! And all of us! All friendly and peaceful. How convenient during the rainy season. A sublime shelter!

The zoologists were so excited about him. Such an opportunity to observe a mature wild Orangutan for a long time from so close is hard to come by. This was their chance.

His serene calmness and magnificent size impressed them. A fully grown Orangutan, be it male or female, is immensely strong. However, no Orangutan is naturally aggressive when left in peace in the wild. He was watching the people as thoroughly as they were scrutinizing him. He did not at all mind being photographed. He is so cool! Such a gentleman.

Talking about strength; let us just tell you about how the zoologists at the station gave some of the bigger apes in quarantine an injection. It was quite a sight! The ape patient was in the cage. Four different people outside the cage each grabbed a foot or a hand through the rails, and held on with all their might. Meanwhile, a fifth person got the injection ready and injected the poor creature with the necessary medicine. It did not last long, but that is what it takes to dominate us.

We loved studying his face; he looked so wise and mature. Probably, he did not run into deforestation and bad treatment from human beings as we did. He seemed to us the essence of a wild magnificent Orangutan at peace. He is lucky. Unfortunately, these days, it appears that we can only survive with the help of human beings who care to protect us.

The more he came to visit us, the more he felt at ease communicating with us and the people from afar. He often had them quite startled while they were walking up the mountain path. He would make this smacking sound from high up in the trees. It sounded so human they could not quite figure out where that noise came from. Then he would throw a stick down, landing precisely just in front of their feet. Who was watching them? Up they looked... and sure enough there he sat looking for attention.

The zoologists took great pleasure in watching us climb. We, the littler ones, goofed around climbing up and down trees, apparently amusing them. Mehra and her friend, however, were a veritable treat to watch, even for us, when they arrived at or departed from the feeding site. They just have to be the most elegant climbers anyone can ever dream of.

Their knowledge of a particular tree's flexibility is awesome. They would reach out with their long arms swinging in order to touch the nearby crown of the adjacent tree. When the distance was right, they let go of the old tree and on they sailed to the next one. It was often hard to see due to the immense height of the trees in the rain forest—up to a stunning 200 feet—but we always heard them crashing through the very top of the rain forest.

The scientists studied to find out how long our life span really is. It was not easy for them, since, when we are wild and free, we are hard to find and too high up to track us down. Therefore many observe us in a zoo. Ugh! Who wants to live in a zoo? It is like being in a prison and the results of their studies were accordingly. But anyhow, they assumed we live for about 40 years on average. If these nice zoologists manage to educate people about the importance of the rain forest and its native population, we have a chance to survive and live full lives.

Remember:

1) *Mehra and her wild boyfriend*
2) *Climbing power*
3) *Observing a real free Orangutan*

Isn't it grand to hang on two hands and one foot to check out the scene?

Let us hang loose!

And looser!

SEPPLI

Hello, may I introduce myself? My name is "Seppli", named after a little farmer boy in Switzerland. I am a bit of a clown and goof head and must have "swooped" the zoologists with my natural charm. Well, I came here confiscated by some most authoritative people. I am young, four years old, and here to learn how to make a life of my own in the forest.

I was always up for any tricks and loved to fool around during the teaparty. The people laughed and laughed and were so entertained. They loved my long blond eyelashes.

I was most anxious to have that teaparty get started. Mostly first in line, I just could not wait for the delicacies the fine people were carrying up the path. So I just grabbed as much as I could and it often got me into trouble. That milk would fly all over the place. Can you see all the white spots on my head and arm?

My goal was to gather a substantial supply of bananas (as you can see here) to a safe spot; then I would devote concentrated effort to appreciate them. It was hard work to eat all these bananas.

Round-bellied and satisfied, here I sit. After the feast, I played with my friends and the people and their small terrier, Wumi. She was a spirited little creature. We could carry her up and down the trees. That frightened the people. They were talking a lot but then again they would be really goofy themselves and we had a lot of fun and happy times with them. Shortly before it got dark, the zoologists put me with my companions into the cages for the night. Too bad, but we fell asleep fast.

I loved the station. There were at least six others my age and we were a good group. The zoologists were really loving and took excellent care of us. We ate well, were comfortably sheltered during the nights and we got to meet bigger Orangutans like our parents. It was a fine solution and I was not yet ready to leave this place.

Sometimes, during the night, one of the bigger Orang-utans came to the cages studying how to open the locks and letting us all out. We are very smart guys and can figure out a lot. The zoologists disapproved because we were not properly protected. Each time it happened, they would think of more complicated locks to do the job. I really do not think that they considered us so cunning and clever and they must have under-estimated our strength. Alas, after many attempts, they managed to apply locks to the cages which we could not pick! They outsmarted us.

In December the rainy season starts. It is called the monsoon. A wet and gloomy time and not much fun! I caught a bad cold; at first just the sniffles and later I got this deep cough. It sounds just like when humans cough. The zoologists gave me extra food, rich in vitamins, and even took me to their house to stay dry. I began to run a high fever and that is when they took me into the big city, Medan. It was a long and weary journey. They brought me to their human medicine man. He treated me as best as he could; I got worse and worse, nothing helped. After one day there, my cold turned into a pneumonia. I was so sick. The people stayed with me and most tenderly cared for me all night, but I did not make it and died.

Remember:

1) *Seppli the trickster*
2) *Lock-pickers*
3) *Seppli catches a cold*

Please do NOT destroy the Rain Forest. It is MY home!

SUMBING

Can you see my hare-lip? It saved my life because, when I was caught, the people did not think I was a pretty enough girl to keep and so I got a ride to the station. It felt good to be in a place where I could just be. I was quite shy, so I never really socialized. I was just happy living at the station. I was older, about six when I got there, and I knew how to sleep up in the trees. The zoologists did not press me to leave. I had time to get well prepared.

Eating all that delicious banana and drinking the milk was no easy task with my hare-lip; it would just dribble right out from under me. Alas, there was nothing I could do and I did not particularly mind.

I stayed close to the station and always looked forward to the teaparty. There was action then. I enjoyed it. The zoologists were interested in my different lip, wondering and theorizing where it came from. I was just born that way, it seems to me.

Remember:

1) Sumbing the hare-lip

THE REST OF US

This is the picture that the zoologists of the station took of me when I was sitting on the roof collecting bananas for my wild friend and me. My name is Mehra and I am the mature female who was given the wonderful chance of living free in pristine rain forest again.

I was so glad to have been taken to the Orangutan Rehabilitation Station. What more do I want, after I narrowly escaped being stuck in captivity or having to deal with other ghastly things. The people at the station were so gentle to me and, since I was an adult, I did not spend a lot of time there.

The station was a pleasant experience and I always enjoyed coming back. The zoologists were thrilled each time I appeared and were most welcoming to my wild friend. We had an excellent relationship as it was clear that I was happy to be wild and they were happy to see me free.

Me? I was in Seppli's gang. We were good buddies and played endless tricks together.

The people of the station took this photograph to show off our special Orangutan nose. It is short and stubby. We do not have the most refined sense of smell. Whenever we need to know the scent of something, we touch the object in question and then put our finger right into our nostrils. That is the best way for us to dissect that odor. This nose action seems to make humans laugh, even though we are not trying to be funny.

Other animals have a very sophisticated sense of smell; I suppose we are more like humans that way.

I loved watching and getting to know the people. They had such fine hair, however only on their heads. They always walked upright. We climb and swing in the trees, but when we are on the ground we walk on all four. We cannot really stand for a long time. Our arms are just about as long as our legs and so we just use them all. We really wanted to find out as much as possible and spent a lot of time figuring these humans out by touching them. They did not mind. We all became good friends and although our language is different from theirs, we communicated well with each other.

Some days the zoologists came with medicine which we had to ingest. Apparently it was stuff which would rid us of possible worms inside. We had to take a good swallow. All these little tubes and measuring cups! So much fun to participate and grab hold of everything possible! The people tried hard to keep this procedure under control and did not want us to prepare our own dosages. What an ordeal it was!

When they brought their cameras is when life began to be most entertaining. They took so many photographs that often times they had to open the camera and put a new roll of film in there. Those rolls of film looked so exciting and we would just long to get hold of one. The zoologists panicked, most probably because we took great pleasure in unrolling them. Apparently a roll of film is not the same after it has been unrolled. But we caught on pretty fast as how to deal with these cameras and film.

Sometimes, we just needed to rest. Teaparty activities tired us smaller ones and were often rough. Here I am taking a break, thinking of what life would be with my Mom. We are all orphans here, but luckily we have each other.

Here are a few of us older ones. We were free day and night but mostly stayed around the station. It was expected of us to go back into the forest and the zoologists encouraged us to leave. They were giving us less food to motivate our departure. But it was not so easy to just take off and leave our friends here. The people began to ignore us more and more.

The zoologists began to think of an alternative. Too many of us hanging around and not taking our life into our own hands resulted in being taken further into the rain forest, far away from the station. This was a complicated procedure.

We would be packed into the big box again and the men would carry us up the river for at least a day. Then, at the designated area, we were released and there we were, on our own! Every so often, one of us would find our way back. The people were thrilled... well, sort of!

When all efforts failed and some of us just would not leave the station, the people organized a trip in a whirly bird machine to take them even further away to a different reserve. This was very costly and it only happened on rare occasions. It was quite impressive to the ones who went. Once gone, it was impossible for us to get back. So we really needed to apply what we learned at the station. The zoologists could not track us down anymore either and we hope that all of us released ones will be free and well for years to come.

We get grumpy too!

Then we wanted to be left alone.

This is me, Oyong, an adolescent male. I was intrigued with how different the people were with each of us. They just loved some, mainly the littler ones because they were more outgoing towards the zoologists.
And then they were downright uninterested in others. To me, they were snappy. I think I was too old, hung around the station too long and seemed to have an expression that irritated them. I can't help the way I look!

Remember:

1) *Our sense of smell*
2) *Rolls of film*
3) *Leaving the Station*

Do you know what I mean?

Perhaps I should have left long ago!

RAIN, RAIN, GO AWAY...BUT DO COME BACK SOME OTHER DAY

Every year the rainy season begins in December and lasts for about three months.

It rains several times every day; the forest becomes dark green and very wet. When it rains, it pours. A very specific rushing sound is heard seconds before a curtain-like downpour of water takes place. There really is not much protection against getting drenched except a solid house.

Nobody likes that time of year very much; nothing dries and mildew grows. The people looked gloomy and wet and tried to dress themselves in all sorts of funny looking water resisting materials, but then they were usually too hot. The temperature does not really cool off. So the warm temperature together with the moisture makes it a bit like a steam bath.

We really do not like that time of year. We have no house to crawl into for shelter and all of us are wet and uncomfortable. Many of us get colds like Seppli did.

Luckily we had the station, the zoologists and their helpers. They always carried up food twice a day, which really was very appreciated by us. It cheered us up.

We looked a mess. Our deep orange, reddish coat became spiky and brown. It was depressing and we were dreaming of the golden times, frolicking in the warm dry sunshine.

Nothing much would move us away from the station and the people were understanding enough to let us hang out. They seemed to be concerned about our health and were glad to have us nearby to observe us daily.

This is what we would have liked to do most of the day; eat. There really was not much else to occupy ourselves with. The bananas tasted so sweet and refreshing.

The ex-rehabilitants who returned were most welcome. They would travel a good distance relying on their memories of the station and were relieved to find it unchanged. A few days rest perked them up enough to go back to their life in the wild.

The zoologists were so excited to see them come back and make use of their station. After all, that is one of the reasons why the station was built: to help us out while still adapting to living free in the rain forest.

The teaparties continued. The people hiked up the jungle path with umbrellas. They looked very practical to us and we would just love to get hold of them to use for ourselves. They were great fun to open.

When we did not have human umbrellas, we would set out to find one of the big jungle leaves. They are a bit smaller but gave us some relief from the rain. Not quite as efficient as umbrellas, but better than nothing.

We were constantly on the lookout for something new and more protective; so I found this big sign which served its purpose adequately.

The river swelled alarmingly. It rushed furiously by the people's house. It was white and brown and swallowed what it could. It looked scary and dangerous. Sometimes it was so bad that the people could not cross it. They just had to wait until it subsided. It did not concern us much since we cannot swim anyhow.

As you can see, it really was a dreary time.

Although we really do not rejoice in the rainy season, we understand that it is essential to our survival. The rain forest and its rain is like a sponge soaking up the much needed water. If only this fact would convince humans NOT to destroy our rain forest. Will they continue until the Earth's climate is changed beyond hope and the sun parches our planet?

While this rapid destruction of the rain forest is happening at an alarming rate, will we be lucky enough to always find zoologists who protect us? We have had a peaceful gentle life in our habitat and deeply hope that we can continue. We do not want to be imprisoned behind these bars.

It is not right to lock us up, export us to zoos, experiment on us for medical research or even kill us. Nor is it right to invade and destroy our habitat. So please learn about us! We are peaceful, intelligent and sophisticated creatures and without us and so many other animals, the world, manipulated by human beings and their ways, will never be the same again. We really believe in the name of all animals and our beloved rain forest, that we and the people belong to the Earth. Let us live together, respect each other and be sensitive to each other's needs. The Earth is ALL we have; we should pamper our glorious planet.

Remember:

1) *The Rainy Season Begins*
2) *Umbrellas!*
3) *We love the Station*

Epilogue

There are many things you can do to assist the efforts of these zoologists and other organizations to protect the Orangutans, other endangered species and the rain forest. Let me introduce you to a few whom you can contact.

The Bohorok Rehabilitation Station is in existence as well as in transition; the Indonesian government took it over and it became extremely sucessful. So succesful, in fact, that it is being overwhelmed by visitors. Up to 500,000 visitors pour into the station annually. The original purpose of rehabilitation is no longer really feasible with this exposure. That station is being turned into the Bohorok Orangutan Centre catering predominantly to ecotourism. It will be an educational centre with plenty of information on the rain forest and its inhabitants. Some rehabilitants will be exhibited for informative and educational purposes only.

Meanwhile a site for a new Orangutan Rehabilitation Station is being sought in an area devoid of native wild Orangutans.

This transition needs big financial support from all possible sources and you are invited to become donors to this necessary project. Please contact or send your donation to:

PanEco Foundation
Chilehus
8415 Berg am Irchel
Switzerland
Tel: 41 52 318 23 23
Fax: 41 52 318 19 06

There is another organization which is focused on the survival and well being of the Orangutan. Any support will help the Orangutan!

Orang Utan Foundation International
822 S.Wellesley Avenue
Los Angeles, CA 90049
Tel: (310) 207-1655
Fax: (310) 207-1556
E-Mail: redape@ns.net
http://www.ns.net/orangutan

For other publications by Bettina Milliken and more Orangutan photos, please visit http://www.roadrunner.com/~bettina/